The LinkedIn Arborist's Playbook

Get real work, hires, or get hired in just 15 minutes a day, and become an authority without the spammy DMs.

"What I wish my younger self knew about using LinkedIn so I wouldn't have wasted so much time!"

By John Snow, BCMA

TREE CHECK UP

DEDICATION

To my son Nick Snow—

May this guide help you build the business and life you deserve. Your future clients are waiting to discover what you already know: that trees deserve an expert who truly cares.

Keep growing, son.

—Dad

TABLE OF CONTENTS

TABLE OF CONTENTS

TABLE OF CONTENTS

DISCLAIMER & LIMITATIONS

This guide provides LinkedIn strategies based on the author's experience. Results will vary based on your market, consistency, and implementation.

The **Connection-First Method™** is a trademark of Tree CheckUp, LLC and ArboristMentoring.com.

No part of this guide guarantees specific business results. LinkedIn's features and algorithms change regularly. The strategies that work today may need adjustment tomorrow. Keep an eye out for future updates to this guide.

Always comply with LinkedIn's User Agreement and Professional Community Policies. The author is not responsible for account restrictions resulting from misuse of these strategies. For the most current LinkedIn best practices, visit linkedin.com/help.

INTRODUCTION

Most arborists waste months on LinkedIn getting zero results. They post tree photos for other arborists, send spammy DMs, and wonder why property managers ignore them.

This guide shows you the **Connection-First Method**™ that helped me go from around 4000 connections to 5954 connections in 2 months. No cold DMs. No cringe. Just real connections that turn into real work.

The **Connection-First Method**™ is a simple, repeatable system for growing your local presence by building relationships before promoting services.

It isn't a rigid program. It's a **proven framework** you can adapt to your schedule. Some weeks you'll do more, some less. The key is consistency, not perfection.

PRO TIP! *While you can read this guide anywhere, I recommend you do the actual LinkedIn work on a desktop for easier navigation.*

WHO THIS IS FOR

You're in the right place if you're:

- A tree service or landscape company owner (solo to multiple crews).
- A green industry consultant building your authority and client base.
- A sales representative who wants your own pipeline.
- A crew leader ready to step into business development.
- Someone seeking better opportunities in the green industry.

This guide works if you:

- Need more commercial clients (HOAs, property managers, municipalities).
- Want to position yourself as the local expert.
- Are tired of competing on price alone.
- Have just 15 minutes daily to invest.
- Value relationships over transactions.

The Connection-First Method™ works for:

- Business owners building their client pipeline.
- Consultants establishing thought leadership.
- Job seekers connecting with the right employers.
- Anyone who needs quality professional relationships.

Note for Landscapers: Every property manager who needs tree work also needs landscape services. Use the same strategies, just adjust your examples.

Note for Job Seekers: Track 2 (Recruiting) works in reverse to get you noticed by the right employers.. Use it to showcase your skills and expertise, connect with company owners, and stand out from the pile of resumes. Your posts about safety, technique, or passion for the work will get you noticed by the right employers.

Note for Consultants: Track 3 (Authority) is your path. Build your recognition as the expert others tag for answers.

THE ARBORIST'S LINKEDIN RESOURCES

We've built an accompanying web page for all of the resources mentioned within this book.

Each resource mentioned has its own link, but for your ease, everything can be found in one place.

(Especially useful if you're reading the printed version of this book.)

Access the Resources page

arboristmentoring.com/book-resources

WHY CREATE THE LINKEDIN ARBORIST'S PLAYBOOK?

I joined LinkedIn on December 16th, 2008. Crazy right?

That's 17 years I've been on LinkedIn.

I got started using LinkedIn mainly because I was managing a local tree care company, and I was hoping to find employees. That's how I began my journey to connect with over 5000 Arborists around the world.

Over the years, LinkedIn's changed a lot, and so have I. I want to share with you some of the best tips I have so you can be more effective starting today.

Why?

Well, I see far too many company owners, sales arborists, and others using LinkedIn completely wrong!

Take this post below...

2 Likes?

It's a good post, communicates something that Arborists all see and understand.

But the question is: **Who's really going to see this?**

Probably the 134 **arborist** connections you're already connected with.

Result? Zero new clients!

Who is the target market for this?

The local homeowner, HOA, or your commercial client.

BUT **they aren't getting the message because you're not connected with them.**

▲ THIS GUIDE ISN'T FOR YOU IF:

- ✕ You want instant results (this takes 30-90 days).
- ✕ You won't commit to 15 minutes daily.
- ✕ You prefer mass messaging over building relationships.
- ✕ You're looking for shortcuts or hacks.

■ THIS GUIDE IS FOR YOU IF:

- ✓ You'll follow the system for at least 30 days.
- ✓ You value quality connections over quantity.
- ✓ You're ready to become a recognized expert.
- ✓ You understand that consistency beats intensity.

If this sounds like you, here's how to get the most out of this playbook.

HOW TO USE THIS GUIDE

1. Pick **one primary track** for the next 30 days: **Local Clients**, **Recruiting**, or **Authority**.

2. Follow the **Connection-First Method**™: Connect → Observe → Interact publicly → Post value → **DM only when they engage.**

3. Post **3×/week** and spend **15 minutes/day** total on LinkedIn activities.

4. **If you're a local company**, make everything **local**: name neighborhoods, ordinances, reviewers, and weather patterns.

5. **Promise:** This is a no-fluff field guide. It's short, practical, and built for arborists.

QUICK START GUIDE

Your First Week on LinkedIn

Stop overthinking. Start doing. Here's your exact roadmap:

⏱ Total Time Investment: 45 Minutes to Launch (then 15 minutes daily)

STEP 1

Profile Emergency Fix (30 minutes - Do THIS First)

- ➡ Fix your headline
- ➡ Update your photo
- ➡ Create your banner
- ➡ Rewrite your About section

☑ Complete this before moving forward

STEP 2

Choose Your Track (2 minutes)

Pick ONE for your first 30 days:

LOCAL CLIENTS
You need more work

RECRUITING
You need more crew members

AUTHORITY
You need more recognition

STEP 3

Your First Actions Today (15 minutes)

SEARCH
"[Your City] property manager" or "[Your City] HOA"

FILTER
People → 2nd degree connections

CLICK
"Connect" on 10 profiles (no message needed yet)

CREATE
Your First Post (5 minutes)

Copy and customize this template:

"After last week's [weather event] in [Your City], I'm seeing a lot of [specific tree problem]. Here's what property managers should check today: [specific tip]. What tree concerns are you seeing at your properties?"

STEP 4

Your Daily 15-Minute Routine (Starting Tomorrow)

Morning (5 min)
Send 5 new connections

Midday (5 min)
Like and comment on 2-3 local posts

Afternoon (5 min)
Create one post (Mon/Wed/Fri only)

YOUR WEEK 1 SUCCESS METRICS

✓ Your profile is completely updated.

✓ You've chosen a track and committed to it.

✓ You've sent 25+ connection requests.

✓ You've published 3 posts (Mon/Wed/Fri).

✓ Your daily 15-minute routine completed 5/7 days.

Common Week 1 Mistakes to Avoid:

✗ Sending connection messages (**just click connect**)

✗ Posting generic tree facts (make it **SPECIFIC + LOCAL**)

✗ Trying all three tracks at once (pick **ONE** to start)

✗ Overthinking your posts (**done beats perfect**)

✗ Skipping days (**consistency matters most**)

IF YOU ONLY REMEMBER THREE THINGS:

1. **Fix your profile FIRST**
   Don't connect with a broken profile.

2. **Local beats global**
   Every post should mention your area or be specific.

3. **No cold DMs**
   Let them engage with your content first.

What Success Looks Like:

- Day 1-3: Feels awkward, no engagement (normal!)
- Day 4-7: First connections accepted, someone likes a post
- Week 2: Prospects start viewing your profile
- Week 3: Your first meaningful comment or message
- Week 4: Your first "Can you help with..." inquiry

READY? YOUR FIRST ACTION!

Stop reading. Start doing.

- Go and fix your profile.
- See you back here in 30 minutes to choose your track.

REMEMBER!

Every successful LinkedIn arborist started with a bad profile and zero engagement. I did too.

The only difference? They started anyway.

Now, some people call me The LinkedIn Arborist.
https://www.linkedin.com/in/johnwsnow/

WEEK 1 CHECKLIST

Name: ______________________________

Date: ______________________________

Before You Begin (Day 0)

- Read the Quick Start Guide
- Block 15 minutes daily in the calendar
- Have your phone ready for a profile photo

WEEK 1 CHECKLIST

Day 1 - Monday

Complete the 30-minute Profile Emergency Fix.

- ☐ Headline updated
- ☐ Professional photo added
- ☐ Banner created
- ☐ About section rewritten
- ☐ The featured section has 1+ link
- ☐ Choose your track:

    • Local Clients • Recruiting • Authority

- ☐ Send five connection requests
- ☐ Write and post your first post (use template)

WEEK 1 CHECKLIST

Day 2 - Tuesday

- [ ] Morning: Send five connections (5 min)
- [ ] Midday: Like/comment on 2 LOCAL posts (5 min)
- [ ] Afternoon: Prep tomorrow's post idea (5 min)

Day 3 - Wednesday

- [ ] Morning: Send five connections (5 min)
- [ ] Midday: Like/comment on 2 LOCAL posts (5 min)
- [ ] Afternoon: Post #2 (5 min)

Day 4 - Thursday

- [ ] Morning: Send five connections (5 min)
- [ ] Check who accepted connections
- [ ] Reply to any comments on your posts
- [ ] Prep Friday's post idea

WEEK 1 CHECKLIST

Day 5 - Friday

- [ ] Morning: Send five connections (5 min)
- [ ] Midday: Like/comment on 2 LOCAL posts (5 min)
- [ ] Afternoon: Post #3 (5 min)
- [ ] Review week's metrics (below)

Weekend - Reflect & Prep

- [ ] Batch write next week's three posts (30 min Sunday)
- [ ] Review which connections are accepted
- [ ] Note any engagement patterns

WEEK 1 CHECKLIST

Metrics Tracker

- Connections sent: _____/25 goal
- Connections accepted: _____
- Posts published: _____/3 goal
- Total post likes: _____
- Total post comments: _____
- Profile views: _____
- Messages received: _____

WEEK 1 CHECKLIST

What Worked This Week:

____

____

____

What to Improve Next Week:

____

____

____

Accountability Check:

Did you complete your 15 minutes daily?

Mon ■ Tue ■ Wed ■ Thu ■ Fri ■

Remember: If you're new to LinkedIn, Week 1 feels awkward. Week 2 gets easier. Week 4 brings results.

Don't quit before the magic happens.

YOUR FIRST 90 DAYS:
WHAT'S REALISTIC

Track 1: Local Clients

MONTH 1
Build connections with 50-100 local clients, property managers, or HOA board members.

MONTH 2
Start seeing saves and comments on your posts from non-arborists.

MONTH 3
Receive 2-5 inbound inquiries from people who've been watching your content.

ONGOING
Replace cold calling with warm conversations.

YOUR FIRST 90 DAYS:
WHAT'S REALISTIC

Track 2: Recruiting

MONTH 1
Connect with 100+ arborists in your region.

MONTH 2
Build a reputation as a company that shares knowledge and opportunities.

MONTH 3
Get 3-5 qualified applicants from your network.

ONGOING
Referrals start coming from your connections.

YOUR FIRST 90 DAYS:
WHAT'S REALISTIC

Track 3: Building Authority

MONTH 1

Join conversations with municipal arborists and industry leaders.

MONTH 2

Get tagged in technical discussions.

MONTH 3

Become a recognized voice in your specialty.

ONGOING

Access to peer knowledge that saves you from costly mistakes.

YOUR FIRST 90 DAYS:
WHAT'S REALISTIC

FACTS

This isn't magic.
You'll get out what you put in.
15 minutes daily = gradual growth.
30 minutes daily = noticeable momentum.

WHAT YOU CAN REALISTICALLY EXPECT

WEEK 1-2
20-30% of connections accepted, 0-2 post engagements

↓

WEEK 3-4
First meaningful comments from non-arborists

↓

MONTH 2
1-3 profile views from property managers weekly

↓

MONTH 3
1-2 "Can you help with..." messages

↓

MONTH 6
1-2 new clients directly from LinkedIn

↓

YEAR 1
4-8 solid clients, $30-50K or more in LinkedIn-sourced revenue

WHY 15 MINUTES ON LINKEDIN BEATS 2 HOURS OF WINDSHIELD TIME

The Math Every Arborist Needs to See

Your Current Reality

Drive 30 minutes each way to an estimate = **1 hour**
Spend 45 minutes on site with a tire-kicker = **45 minutes**
Drive back = **30 minutes**

Total: **2 hours 15 minutes**

Close rate on cold estimates: **Maybe 20%. 40% if you're good.**

The LinkedIn Reality

15 minutes connecting with clients.
They see your value posts for 2-3 weeks.
They reach out when they NEED you.
You quote via photos/video first.
Only drive to pre-qualified opportunities.

Close rate: **60%+ (they already trust you).**

WHAT 15 MINUTES ACTUALLY LOOKS LIKE

WHILE YOUR COFFEE BREWS (5 MIN):
Sent five connection requests to local PMs

WHILE WAITING FOR THE CREW TO LOAD UP (5 MIN):
Posted one tip with a job site photo

WHILE SITTING IN THE PICKUP LINE (5 MIN):
Replied to 2 or 3 comments

THAT'S IT.
15 minutes of relationship building that actually works.

THE HIDDEN TIME YOU'RE ALREADY WASTING

Scrolling Facebook: 45 min/day average

Texting photos to customers: 20 min/day

Explaining the same thing over and over: 30 min/day

LinkedIn REPLACES this wasted time with activities that compound:

Post once = 100 PMs see it

Answer once = Save as a template

Connect once = Relationship for years

THE CONNECTION-FIRST METHOD™

Don't send cold DMs. Do earn conversations.

1. **Connect**
   Just click connect - no notes needed.

2. **Observe**
   Watch their posts; don't force comments.

3. **Interact publicly**
   Comment only when you have something valuable.

4. **Post Value**
   Local, specific, save-worthy.

5. **Direct Message**
   Only after they engage (like, comment, or reply to your post).

THE CONNECTION-FIRST METHOD™

What to Post so People Want to Talk

- Local tips tied to ordinances and reviewer expectations.
- Board-ready or tailgate-ready one-pagers.
- Mini-cases with measurable results (time saved, mitigation reduced).

What to Avoid

- ✕ Cold pitch DMs, gore, vague before/afters, addresses.

PRO TIP!

Before you start posting or connecting, you need a profile that makes the right people pay attention. That starts here.

WEEK 0 — PROFILE EMERGENCY FIX (Before You Do Anything Else)

➡️ **Time Required:** 30 minutes

➡️ **Result:** A profile that makes locals want to connect, not click away

The Brutal Truth Check

Pull up your profile on mobile (most people will see you there). If you see any of these, fix them now:

❌ Chainsaw selfie or crew group photo where you can't identify anyone

❌ "Owner at [Company Name]" with zero context about what you do

❌ "25+ years experience" (everyone says this)

❌ Banner that's a logo, truck photo, or the LinkedIn default

❌ About section that lists services like a Yellow Pages ad

❌ No location specifics (just "Florida" tells them nothing)

WEEK 0 — PROFILE EMERGENCY FIX (Before You Do Anything Else)

The 30-Minute Fix Formula

1) Headline Surgery (5 minutes)

Replace your current headline based on your track.

Local Clients Track

❌ ISA Certified Arborist | Tree Service | Removal & Pruning

■ Helping [City] HOAs Reduce Tree Risk & Costs | Board-Ready Reports

Recruiting Track

❌ Hiring Tree Climbers

■ Growing Safe Tree Climbers in [City] | Gear, Training, Career Path | Now Hiring

Authority Track

❌ Tree Expert | 25 Years Experience

■ The Arborist [City] Property Managers Call for Second Opinions

WEEK 0 — PROFILE EMERGENCY FIX (Before You Do Anything Else)

The 30-Minute Fix Formula

2) Photo Reality Check (5 minutes)

Your photo needs to say, "I'm the professional you call when trees matter."

Take a new phone photo right now:

- Outside, natural light (The golden hour is perfect)
- Clean shirt with collar (A company polo is fine)
- Slight smile (Be approachable, not goofy)
- No sunglasses; no equipment blocking your face

Can't take one today? Use your newest professional photo where you look approachable. Change it within 7 days.

WEEK 0 — PROFILE EMERGENCY FIX (Before You Do Anything Else)

The 30-Minute Fix Formula

3) A Banner That Sells Your Value (10 minutes)

Your banner is free billboard space. Stop wasting it.

Quick Canva ideas (pick one):

- A photo of before/after of proper pruning (not removals).
- A photo of your crew doing safe, clean work.
- Simple text: "Fewer Tree Headaches for [City] HOAs"
- A photo with your local skyline with a short value prop.

No Canva skills? Use your phone's built-in photo editor to add simple text to a photo of a perfectly pruned tree, then upload to LinkedIn.

WEEK 0 — PROFILE EMERGENCY FIX (Before You Do Anything Else)

The 30-Minute Fix Formula

4) Your About Section: Delete Everything and Start Over (10 minutes)

Formula: Problem you solve + Who you solve it for + Proof + Next step

Template — For The Local Clients Track

"HOA boards and property managers in [City/County] call us when:

- They're tired of 'remove it' being the only recommendation
- Tree work keeps getting kicked back by reviewers
- They need someone who speaks both arborist and board meeting

We provide board-ready reports, reviewer-friendly documentation, and specifications that actually protect trees while reducing liability."

WEEK 0 — PROFILE EMERGENCY FIX (Before You Do Anything Else)

Recent Win

[Neighborhood] HOA was quoted $120K for tree removals. We got it down to $72K by saving healthy trees they thought were hazardous.

> Let's connect if you manage trees in [County]. I share weekly local compliance tips and board-ready checklists."

Template — For The Recruiting Track

"We're building the next generation of arborists in [City]."

If you're new: You'll learn knots, rigging, and safety on the ground before you climb. We provide gear, training, and a clear path to certifi cations.

If you're experienced: You'll work with better equipment and a crew that has your back.

What makes us different:

✓ We pay $18–$35/hr based on experience
✓ Weekly safety training that's actually useful
✓ ISA certification support (we pay for tests)
✓ No drama, home every night

→ Connect to follow our crew stories and openings. We're hiring two climbers this month."

WEEK 0 — PROFILE EMERGENCY FIX (Before You Do Anything Else)

5) Your Featured Section Setup (5 minutes)

Pin these in order:

1. Link to your best value (HOA checklist, hiring process, etc.)
2. Newsletter signup (if you have one)
3. Your most-engaged LinkedIn post (social proof)

No content yet? Pin:

- Your company website.
- A news article featuring your work.
- Your ISA/TCIA certification lookup

WEEK 0 — PROFILE EMERGENCY FIX (Before You Do Anything Else)

The "After" Checklist

Before moving to Day 1, verify:

- [ ] Headline says who you help and how.
- [ ] The photo shows a professional that the client would trust with their
- [ ] trees.
- [ ] Your banner reinforces your value (not just pretty trees).
- [ ] The About section makes you the ideal person the client wants to connect with.
- [ ] Your location is specific (city/county, not just state or United States).

Your featured section has at least one valuable link.

WEEK 0 — PROFILE EMERGENCY FIX (Before You Do Anything Else)

Common Excuses (and Why They're Wrong)

"I don't want to put myself out there."	Your competitors already are. Be findable or be forgotten.
"My clients aren't on LinkedIn."	Their bosses are. Property Management firms, HOA managers, and other decision makers live here.
"I hate writing about myself."	Use the templates. Copy, paste, edit locations. Done.
"What if other tree services copy me?"	They can copy words, not your reputation, relationships, or results.

Day 1 starts when this is done.
Don't send a single connection request with a broken profile.

CHOOSE YOUR TRACK (30-DAY FOCUS)

PRO TIP! Now that your profile is dialed in, the next step is choosing the track that matches where you want to grow.

Pick one. You can rotate next month.

Local Clients (HOAs/condos, property managers, developers, municipalities, homeowners)

Recruiting (climbers, crew leads, sales arborists, trainees)

Authority (be the person people tag for tree questions in your area)

Decision rule: If you have a **revenue gap**, choose *Local Clients*. If you have a **capacity gap**, choose *Recruiting*. If you need **brand pull**, choose *Authority.*

TRACK 1 — LOCAL CLIENTS: Find & Connect (No Cold DMs)

Start with the Connection-Only Workflow

Step 1: Connect First (No Notes Needed)

Search for
local property managers
and HOA board members

Click connect
No message required

Let them
ACCEPT or IGNORE

TARGET 50
YOUR LOCAL LIST (WORKSHEET)

Goal: Connect with 50 local decision makers who can hire or refer you.

The Tracking Trap I Escaped (So You Don't Have To)

I used to maintain massive spreadsheets:

- Hundreds of names
- Connection request dates
- Response tracking
- DM histories
- Follow-up schedules

Total waste of time.

Here's what I learned after tracking 500+ connections: **The only people who matter are the ones who engage.** Period.

That spreadsheet with 500 names? Maybe 10 became clients. And guess what? Those 10 all engaged with my content first. I could have skipped the other 490 entries. But having those 490 also adds value. You don't know who they know (possible referral) or when they might have an interest or need. Keep them too.

The Connection-First Method™ is simple:

- Search for local professionals.
- Click connect (no messages, no tracking).
- Post valuable content.
- When someone likes, comments, or saves → THEY just raised their hand.
- NOW they're worth your attention.

No spreadsheets. No CRM. No tracking apps.

Just connections, content, and conversations with people who actually care.

Note:

LinkedIn's free version no longer offers tagging. If someone engages repeatedly, just remember their name or use a simple notebook. The people who matter will make themselves obvious through their engagement.

Focus on engagement, not administration.

BONUS

YOUR HIDDEN LINKEDIN GOLDMINE

The Overlooked Strategy That Changes Everything

You're sitting on a goldmine - your existing client database.

These people already trust you, and their LinkedIn networks are full of similar prospects.

Why This Works:

- These people already know and trust you.
- They're proof that you work in their area.
- Their connections are pre-qualified leads.
- Zero cold outreach needed.

The 5-Minute Client Connection Process

1. List your best 20 clients (start small).
2. Search for each on LinkedIn.
3. Send a simple connection request.
4. They'll likely accept - they know you!

After They Connect:

Your existing clients become your best advocates when they:

- See, like, and share your helpful posts.
- Comment on your content (Their comments are visible to their network).
- Tag you in relevant discussions.
- Introduce you to their connections.
- Become your LinkedIn advocates.

The Natural Network Effect

When you post valuable content:

- Your clients engage with it.
- Their connections see this engagement.
- Their connections might check out your profile.
- They might request to connect (or you can request - you share a mutual connection).

Smart Connections to Look For

Within your clients' networks, watch for:

- Other property managers in their area.
- HOA board members (they all know each other).
- Vendors who service multiple properties.
- Municipal contacts.

You're reconnecting with people who already know **your work** and naturally expanding to their **trusted circles**. It's how business has always **worked**, just on LinkedIn.

This strategy alone could literally double your connection quality while halving the time to build trust!

TRACK 1 — LOCAL CLIENTS: Find & Connect (No Cold DMs)

The Connection-Only Workflow

Step 1.5: After They Accept

Skim their posts ONCE or TWICE a week

Comment when you have something valuable to add

Otherwise, JUST REACT (like, insightful, support) to stay visible

This shows you're paying ATTENTION without being pushy

The Connection-Only Workflow

Step 2: Build Relationships Through Value

- Post locally relevant content 3x/week (use formulas below).
- Check your connections' posts 1-2x/week.
- Like posts to stay visible.
- Comment **ONLY** when you can add **real value.**

TRACK 1 — LOCAL CLIENTS: Find & Connect (No Cold DMs)

The Connection-Only Workflow

Step 3: Let Them Come to You

Two scenarios for DMs - 1 of 2

When they comment on your posts:

- Reply **thoughtfully** to their comment publicly first.
- Then send a **casual DM** continuing the conversation.
- Example: "Good point about the insurance requirements. How long have you been managing properties in [area]?"
- Let the conversation **develop naturally** over 2-3 exchanges.
- **Only mention** resources if they fit the flow.

TRACK 1 — LOCAL CLIENTS: Find & Connect (No Cold DMs)

The Connection-Only Workflow

Two scenarios for DMs - 2 of 2

When they like multiple posts:

- Simple acknowledgment: "Hi [Name], noticed you've been. following my posts. How long have you been managing properties in [area]?"
- Keep it conversational - you're starting a dialogue, not a sales pitch.
- Let them guide where the conversation goes.

TRACK 1 — LOCAL CLIENTS: Find & Connect (No Cold DMs)

The Connection-Only Workflow

The 2-3 Message Rule

Exchange at least 2-3 messages before mentioning any resources.

Focus on:

- Learning about their specific challenges.
- Sharing a relevant quick tip or insight.
- Building actual rapport.

REMEMBER!

The goal is relationship first, resource second. If they never ask for the checklist, that's fine. You've still built a connection with someone who might need tree work later.

YOUR CONTENT STRATEGY:
WRITE POSTS THAT SPARK *THEIR* ENGAGEMENT

The Local Post Formula That Actually Works:

Formula #1: The Storm/Event Response

[Recent Event] + [Specific Location] + [What to Check] + [Visual Proof]

Example:

"That storm last night hit [Neighborhood] hard. Here's what to check on your trees today [3 photos of actual damage]"

Why it works:

Timely, local, helpful, and proves you were there.

Formula #2: The Problem Spotter

"Seeing a lot of [specific issue] in [area] this [timeframe].
Here's [early warning sign] that [benefit]"

Example:

"I've been seeing a lot of oak wilt in [North Arlington] this month. Here's the early warning sign that saves thousands."

Why it works:

It positions you as the local expert who notices patterns.

Formula #3: The Money Saver

> "Helped [Location/Client Type] avoid [big cost].
>
> Sometimes the answer is [simple solution]"

Example:

"Helped [Riverside] HOA avoid a $50K removal. Sometimes the answer is proper pruning."

Why it works:

Specific savings + challenges assumptions.

Formula #4: The Municipal Update

"[City] just [changed something]. Here's what [audience] needs to know."

Example:

"[Dallas] just changed their tree permit process. Here's what property managers need to know."

Why it works:

Breaking news, they probably missed.

Formula #5: The Myth Buster

> "[City]'s most [superlative] tree isn't [expected]. Here's what we're actually [doing] most"

Example:

"[Austin]'s most dangerous tree isn't what you think. Here's what we're removing most after storms."

Why it works:

Challenges assumptions with local data.

Pro Tip: The Key Patterns for Good Posts

- Always include the specific neighborhood / area.
- Add photos whenever possible.
- Use real numbers from actual jobs.
- End with a question to spark comments.
- Always tie back to their actual problems.
- Write like you're texting a friend about work.
- Post within 24-48 hours of local events.

The Golden Rule:

If you wouldn't text it to a client, don't post it.

THE ARBORIST'S GUIDE TO SAVEABLE CONTENT - GETTING YOUR POSTS SAVED!

What Clients Actually Save

🏆 Top Savers (15-30+ saves):

- Municipal requirement checklists by city 27
- "Common rejection reasons" lists
- Tree mitigation cost calculators
- Species-specifi c care calendars
- Storm prep timelines
- Budget planning templates

■ Solid Savers (5-15 saves):

- Before / after case studies with costs
- "Questions to ask your arborist" lists
- Seasonal maintenance schedules
- Code change summaries
- Risk assessment examples

THE ARBORIST'S GUIDE TO SAVEABLE CONTENT - GETTING YOUR POSTS SAVED!

📌 **Rarely Saved (0-5 saves):**

- Company announcements (unless it's your competitor)
- Industry news shares
- "Happy Friday" posts
- Generic tips without specifics
- Opinion pieces

The Save-Worthy Formula

Reference Value + Specific Numbers + Local Relevance = Saved Post

Example Transform:

❌ "Tree care is important for property value."

⬛ "Tampa HOAs: 2024 Tree Requirements Checklist **(Save this for your next meeting)"**

TRACK 2 — RECRUITING:
CONNECT WITH ARBORISTS & FUTURE EMPLOYEES

LinkedIn started out essentially as a way for employers and employees to connect. Since then, it's evolved to be much more, but the core of the idea is still there, and you can take advantage of it.

Think about it, LinkedIn isn't going to solve that "I need a new climber tomorrow," but it can help you find employees that will stick around.

Why's that? Simple:

LinkedIn, at its very core, is about not just making connections with people but actually developing relationships over time.

When you connect with other arborists, get to know them; they may, in fact, be that future employee or can refer one to you.

I've seen too many ads on LinkedIn for a groundsperson, climber, or municipal employee. Sure, that might work sometimes, but it's basically like shouting out your job listing to the universe, and it's going to get lost.

Which is better?

A future employee who just saw your ad or someone you already know, connected with, and had a couple of conversations with, who's interested in joining your team ?

Who would be your first choice for the next employee:

✕ Random person who saw your ad.
■ Someone you've built a relationship with.

People want to work with **people they know, trust, and look up to**. LinkedIn is your opportunity to set **that stage.** If you know how ! Don't expect to hire someone next week.

This is about building a talent pipeline over 3-6 months.

How To Find Candidates (No Cold DMs)

Watch This First: How to Find and Connect with Arborists

https://treecheckup.com/find-arborists-video

In this 3-minute video, I show you exactly how to search for qualified arborists in your area and connect with them the right way.

After Watching the Video, Here's Your Search Formula

In the Search Bar start with:

[job titles] + [certifications] + [your location]

Job titles: arborist, climber, groundsman, tree worker

Certifications: ISA, TRAQ, TCIA (shows they're serious)

Your cities: Add 3-5 cities within your hiring radius

Example Search:

If you're in Tampa, try combinations like:

- ISA certified arborist + Tampa.
- Tree climber + St. Petersburg.

PRO TIP!

- LinkedIn's search isn't perfect.
- Try different combinations and save the searches that give you the best results.
- Also, remember, LinkedIn is a professional network, so you're much more likely to find a certified arborist or experienced climber than an entry-level person.

Advanced Search Tips

Start Simple:

- Search "Certified Arborist" + your state
- Focus on 2nd-degree connections (they accept faster)
- Just hit Connect - no message needed
- Do 10-20 per session, then rotate search terms

Rotate Your Searches

Keep it fresh by alternating between:

- Certified Arborist
- Urban Forester
- City Arborist
- ISA Certified
- Tree Climber

Stay Organized

- Track who accepts
- Add them to a "Recruiting" list
- Note which companies they work for
- Watch for their engagement on your posts

REMEMBER!

This is a numbers game at first.

Connect with lots of qualified people, **post** valuable content, and let them **come to you** when they're **ready**.

Other searches you can perform include:

- LinkedIn **Groups**:
  ISA state chapters, local trade groups → connect with active posters.
- **Followers of** local tree companies → connect with active commenters.
- **Schools / Alumni**: Forestry/landscape programs near you.

Write Posts That Invite Engagement 2-3x a Week

- "**Train-to-Climb**: Your first 90 days (skills + safety path)."
- "Climbers start at $25-35/hour here in Tampa."
- "Here's what certified arborists make at our company."
- "**Our pay scale**: Groundsman to Lead Climber."
- "**3 pre-climb checks we never skip** (photo + checklist)."

DM Only After They Engage

"Thanks for the comment! Want me to send you our 90-day training path?"

Hey, saw you liked my post. Happy to share our pay scale + gear list if you're interested."

Offer a Regular Monthly Referral Engine Post

For Example: "Adding 1-2 climbers in [City]. Train-to-Climb path, paid certs, no drama. Know someone solid? Comment **JOIN**. Referrals that stick 90 days = **$500**."

The specific pay info works great in posts because it:

- Gets attention
- Shows transparency
- Attracts serious candidates
- Filters out people looking for different pay ranges (so you don't waste your time interviewing them)

TRACK 3 — AUTHORITY: BECOME THE LOCAL TREE EXPERT

Ultimately, as you grow with LinkedIn, this is the pinnacle of where you want to be. While you might be focused on recruiting or connecting with potential clients, in the end, it's all about what you deliver.

This is how you do it.

Here's the test:

Think about who gets tagged when someone has a tree question in your area.

Is it you? Or your competitor?

Authority isn't about having the most certifications or years of experience.

It's about being the person people think of **FIRST** when trees matter.

When you're the recognized expert:

- Clients seek you out (no more chasing).
- You can charge premium rates.
- Competitors become referral sources.
- Media calls you for quotes.

This track transforms you from **"another tree service"** to **"THE tree expert"** in your market.

Choose Authority if you:

- ■ Have solid business but want premium clients.
- ■ Want to influence local tree policies.
- ■ Are tired of competing on price.
- ■ Want media recognition.
- ■ Plan to expand into consulting (Like I did).

AUTHORITY BUILDING WORKFLOW

Step 1: Position Yourself as the Expert (Not Just Another Service)

Week 1: Find Your Authority Angle

Define your authority angle: "**The [City]** arborist who **[unique expertise].**"

But make it real. What specific problem do you solve better than anyone else in your area?

For example:

- "I'm the guy Tampa developers call when their tree permits get rejected."
- "I help Orlando HOAs prep for hurricane season without overspending."
- "I save historic trees that other arborists want to remove."

Or even simpler:

- "The permit guy."
- "The storm prep specialist."
- "The tree saver."

> Your angle emerges from what people already ask you about most.

Week 2-4: Build Your Knowledge Repository

- Create **3-5 "reference posts"** that people will save.
- **Focus on technical knowledge** simplified for decision-makers.
- Always **tie back** to local regulations or conditions.

Step 2: The Authority Content Strategy

Monday - Educational Deep Dives

- "Why [City]'s Tree Ordinance Section 4.2 Actually Helps Property Values."
- "The Science Behind Why Topping Fails (With Local Examples)."
- "How [County] Calculates Tree Mitigation (The Real Formula)."

Wednesday - Myth-Busting with Data

- "The $50K Mistake I See Weekly in [City]."
- "Why Your Neighbor's Tree Advice is Probably Wrong."
- "3 Tree Myths Even Contractors Believe."

Friday - Industry Leadership

- "What I Told the [City] Council About Tree Policy."
- "Behind the Scenes: Writing the State Model Ordinance."
- "Why I Disagree with the Standard Approach to [Topic]."

Step 3: Engagement Tactics for Authority

Get Tagged as the Expert

Answer Technical Questions Thoroughly

- When someone asks about trees, provide THE definitive answer.
- Include citations, code references, best practices.
- Others will start tagging you: "Ask @JohnSnow, he's the expert."

Create "Tag-Worthy" Content

- Comprehensive guides people reference
- Calculators or formulas
- Decision trees (pun intended)
- Municipal requirement checklists

The Authority Multiplier Move

- Connect with other local experts (lawyers, engineers, architects)
- Comment on their posts with tree-related insights
- They'll start referring tree questions to you

Step 4: Authority-Building Connections

Who to Connect With:

- Municipal arborists and reviewers.
- Environmental consultants.
- Civil engineers and landscape architects.
- Local government officials.
- Industry association leaders.
- Media contacts
  (reporters who cover development/environment).

YOUR AUTHORITY DM
(After They Engage)

"Thanks for the question about [topic].
I actually wrote a guide on this for the Florida Urban Forestry Council.
Want me to send you the link?"

Success Metrics for Authority

Month 1 Targets

- 5+ saves per EDUCATIONAL post
- 2+ technical questions ANSWERED publicly
- 10+ connections with industry professionals

Month 3 Targets

Tagged in discussions 2-3x/week

Invited to speak at local events

Media inquiry or quote request

Competitors asking your opinion

Month 6 Targets

- Recognized at industry events
- Consulting opportunities arise
- Premium clients seek you out
- Become the "go-to" for specific expertise

THE AUTHORITY POST FORMULA

Structure

1. Bold Claim or Surprising Fact
2. Local Proof/Example
3. Technical Explanation (Simplified)
4. What This Means for [Audience]
5. Invitation to Learn More

Example

"90% of tree failures I investigate could have been prevented.

Last week in Westshore: A 40-year-old oak fell on a home. The cause? Improper pruning 5 years ago created a weak point.

Here's what happened: [Technical but simple explanation].

For property managers: This is why your tree **specs** matter more **than** your tree **budget**.

"What tree maintenance specs does your property follow?"

Once you've chosen your track, you're ready for the core system that drives the entire Connection-First Method™. These are the steps you'll use every day. You don't need to use every tool at once—pick what supports your track and your schedule.

NEXT STEPS: YOUR UNIVERSAL LINKEDIN TOOLKIT

The daily actions are the same regardless of which track you picked. Whether you're hunting for clients, climbers, or credibility, you'll use the exact same:

- 15-minute daily routine
- 3 posts per week schedule
- Hook formulas
- Content calendar

The only difference? Who you connect with.

Track 1: Property managers and HOAs

Track 2: Arborists and crew members

Track 3: Industry leaders and municipal contacts

Everything else stays constant.

What follows are the universal tools that work for any track. These aren't track-specific strategies. They're the foundation that makes ANY track successful.

Ready? Let's build your LinkedIn habit.

THE DAILY 15-MINUTE ROUTINE

(YOUR LINKEDIN POWER HOUR IN A QUARTER HOUR)

The Myth-Buster

"I don't have time for LinkedIn" is what arborists say while spending 2 hours driving to estimates that don't close. These 15 minutes will build more pipeline than a week of windshield time.

Why 15 Minutes Works

- Consistency beats intensity - Daily touches > weekend marathons
- LinkedIn rewards regularity - The algorithm loves daily users
- Compounds fast - 15 mins × 30 days = 150 new connections + 12 posts
- Fits anywhere - Coffee brewing, crew loading, lunch break

"But I Hate Social Media!"

Why LinkedIn Is Different

LinkedIn isn't Facebook. No pictures of your lunch. No political rants. No cat videos.

LinkedIn is a professional referral network that happens to live online. Think of it as:

- A 24/7 chamber of commerce meeting
- Your digital business card collection
- Word-of-mouth marketing on steroids

The Arborist's LinkedIn Reality

- Your competitors are NOT here (advantage: you)
- Property managers ARE here (they check it at work)
- No angry homeowners complaining about prices
- Professional discussions only

Still skeptical?

Your Non-Negotiable Daily Actions:

☕ 5 Minutes: Morning Connections (While Coffee Brews)

- Search your metro for property managers/HOA boards
- Send five connection requests (no notes!)
- Add them to your Target 50 list
- Why mornings? People accept connections during their inbox check

Your Non-Negotiable Daily Actions:

3 Minutes: Quick Engagement (While Crew Loads)

- Skim your feed for LOCAL posts
- Drop 1-3 helpful comments (ONLY if you add value)
- Like posts from your target market
- Why? Engagement tells LinkedIn you're active

5 Minutes: Content Creation (3× per week)

- Monday: Share a local tip
- Wednesday: Post a mini-case
- Friday: Behind-the-scenes story
- Why 3×? Enough to stay visible, not so much that you burn out

The 30-Day Compound Effect

- Day 1-7: Feels like shouting into the void
- Day 8-14: Connections start accepting
- Day 15-21: People recognize your name
- Day 22-30: First inbound message arrives

Pro Scheduling Hacks

- **Set a phone timer** - When it dings, stop (perfectionism kills consistency)
- **Batch on Sunday** - Write all three posts for the week
- Use downtime - Waiting at the parts store? Send connections
- **Track streaks** - Mark an X on your calendar for each day completed

What This Replaces

Instead of:

✗		✓
Scrolling Facebook for 45 minutes	>	Build your LinkedIn network
Explaining the same thing repeatedly	>	Post it once, send the link
Cold calling 20 property managers	>	Connect with 100 this month

The Bottom Line

In the time it takes to drink your morning coffee and get the crew rolling, you've done more business development than most tree services do all week. That's how you win.

THE 3-POST WEEKLY SCHEDULE

Why 3 Posts Per Week Works Better Than Daily:

The LinkedIn Algorithm Reality:

- LinkedIn rewards consistency over frequency.
- 3 quality posts outperform 7 mediocre ones.
- Your audience needs time to engage with each post.
- Posts have a 48-72 hour engagement window.

The Arborist's Reality:

- Monday: Fresh start energy, people checking LinkedIn after the weekend.
- Wednesday: Mid-week engagement peak, breaking up the work week.
- Friday: Reflection mode, people are more likely to save/share.
- Leaves Tuesday/Thursday for engagement and relationship-building.

THE 3-POST FRAMEWORK:

Monday - Educational/Problem-Solving

- Share technical knowledge
- Address common problems
- Provide solutions
- Example: "Why your oaks are struggling after construction"

Wednesday - Case Study/Results

- Show your work
- Share wins (yours or clients')
- Demonstrate expertise
- Example: "Saved Riverside HOA $45K with this one assessment"

Friday - Personal/Conversational

- Behind the scenes
- Team highlights
- Industry observations
- Example: "What 20 years of storm cleanup taught me"

WHAT TO DO ON NON-POSTING DAYS:

- Tuesday/Thursday: Focus on engagement.
- Like and comment on others' posts.
- Respond to comments on your posts.
- Send connection requests.
- No pressure to create content.

BATCH CREATION STRATEGY:

- Sunday: Write all 3 posts (30 minutes - 1 hour)
- Monday AM: Post and engage (5 minutes)
- Wednesday AM: Post and engage (5 minutes)
- Friday AM: Post and engage (5 minutes)

THE COMPOUND EFFECT:

- Week 1: 3 posts = 300-500 views
- Month 1: 12 posts = 2,000-5,000 views
- Month 3: 36 posts = 10,000+ views and growing

HOOKS & CTAS THAT ATTRACT ENGAGEMENT

12 Hook Starters (Local + Useful)

1) "Before hurricane season in [City]…"
2) "If your HOA gets 'remove or else' letters…"
3) "What our reviewers in [County] actually look for…"
4) "The pruning mistake that creates more risk than it removes…"
5) "Board agenda ready: 7 questions to ask your tree vendor."
6) "Three photos that save a week on permits in [City]."
7) "How to avoid mitigation surprises in [County]."
8) "When lion's tailing shows up in bids—do this."
9) "A fast way to cut lift time by 2 hours on site."
10) "The homeowner myth I hear every spring…"
11) "Where we actually measure DBH (and why reviewers care)."
12) "If a storm's inbound, this is your 24-hour tree checklist."

HOOKS & CTAS THAT ATTRACT ENGAGEMENT

6 CTA Patterns (use sparingly)

1) "Comment CHECKLIST for the board one-pager."
2) "Comment JOIN for our 90-day path."
3) "Want the printable version? Reply PDF."
4) "Save this for your next board packet."
5) "Share with the PM who handles your trees."

THE 30-DAY POSTING CALENDAR

(Your Content Roadmap to Consistency)

Why Most Arborists Fail at LinkedIn

They post randomly when "inspired" → Go silent for weeks → Wonder why nothing happens.

Consistency beats creativity. A mediocre post every **Monday, Wednesday, and Friday** beats a brilliant post once a month.

The 3-2-1 Content Formula

- 3 posts per week (sustainable pace)
- 2 value posts (tips, insights, education)
- 1 story post (builds connection)

Your 30-Day Content Blueprint

Week 1: Establish Your Expertise

Monday:
Local tree tip (seasonal, specific to your area)

Wednesday:
Mini-case from recent work (no client names)

Friday:
Behind-the-scenes (crew photo, equipment, process)

Week 2: Build Authority

Monday:
Municipal requirement update or reminder

Wednesday:
Cost-saving tip for property managers

Friday:
"Myth vs Reality" post about tree care

Your 30-Day Content Blueprint

Week 3: Create Connection

Monday:	Wednesday:	Friday:
Client success story (with permission)	Industry trend affecting local properties	Your "why I became an arborist" story

Week 4: Drive Action

Monday:	Wednesday:	Friday:
Seasonal checklist or reminder	"3 Signs You Need an Arborist" post	Open-ended question to your network

Content Batch Day (Sunday, 30 minutes):

- Review your week's photos/jobs
- Pick your three topics
- Write all three posts
- Schedule or save as drafts
- Done for the week

Never Run Out of Ideas:

Monday Posts (Educational):

- Storm prep tips
- Species spotlights
- Seasonal care reminders
- Municipal updates
- Pruning best practices

Never Run Out of Ideas:

Wednesday Posts (Problem-Solving):

- Mini-cases
- Cost comparisons
- Time-saving solutions
- Common mistakes to avoid
- "What we found" discoveries

Friday Posts (Personal):

- Team highlights
- Project photos
- Client testimonials
- Industry observations
- Your expertise journey

Never Run Out of Ideas:

PRO TIP!

Want your crew to help create authentic content?

Download the free 'Five-Minute Crew Content Playbook'.
https://arboristmentoring.com/wp-content/uploads/2025/11/The-5-Minute-Crew-Content-Playbook.pdf
They can share job site insights while you focus on strategy.

The Emergency Content Bank:

Keep five evergreen posts ready for busy weeks:

- "Top 3 Tree Myths in [Your City]"
- "When to Call an Arborist" checklist
- Your credentials/certifications post
- Seasonal tree care calendar
- A favorite project transformation

Your Actual 30-Day Calendar:

Week	Monday	Wednesday	Friday
1	Local tip: [Topic]	Mini-case: [Project]	Behind the scenes: [What]
2	Requirement: [Which]	Save money: [How]	Myth: [What]
3	Success: [Client Type]	Trend: [What]	My story: [Angle]
4	Checklist: [Season]	3 Signs: [What]	Question: [Ask what]

Tracking What Works:

After each post, note:

- Saves (**your golden metric**)
- Meaningful comments
- DMs received
- Local engagement

<u>After 30 days</u>: Double down on what got *saved*, ditch what didn't.

The Non-Negotiable Rule:

- If you miss a day, don't try to **"catch up"** with two posts tomorrow.
- Just get back on schedule.
- Consistency beats perfection.

Pro Calendar Hacks:

- **Use your phone's notes app** - Draft while waiting
- **Photo everything** - Today's job site is next week's content
- **Repurpose winners** - Your best posts can run quarterly
- **Stay local** - Every post should mention your service area

The Compound Effect:

- **Month 1:** Building habit, finding voice
- **Month 2:** Engagement increases, style emerges
- **Month 3:** Inbound messages start
- **Month 6:** "I see you everywhere on LinkedIn"
- **Month 12:** Recognized expert in your market

Remember:
Your competition is posting "Happy Friday!" with a stock photo. You're sharing valuable insights. That's how you win.

Now that you've got the core tools in place, it's time to put them into practice. For the next four weeks, you'll get one simple challenge that builds momentum fast.

QUICK WIN OF THE WEEK (YOUR WEEKLY LINKEDIN CHALLENGE)

Week 1: Profile Photo Upgrade

- Take a professional photo outside in natural light
- Upload within 24 hours
- Watch your acceptance rate increase

Week 2: Local Hook Master

- Add your city name to every post this week
- Track which posts get the most local engagement

Week 3: The Save Test

- Create one reference post (checklist, guide, or calculator)
- Aim for 10+ saves

Week 4: Connection Quality Check

- Review your last 50 connections
- What percentage are local decision makers?
- Adjust your targeting accordingly

As you build consistency, the next step is creating posts that show your value clearly. That's where mini-cases come in—they give your audience proof, not hype.

PROOF, NOT HYPE

THE MINI-CASE FORMULA THAT SELLS

The Harsh Reality:

Every tree service claims they're "the best" with "25+ years experience" and "competitive prices." Property managers' eyes glaze over. They've heard it all.

What actually gets their attention? Specific proof.

Why Mini-Cases Work:

- Numbers beat adjectives - "Saved 40%" > "We save money"
- Specifics build trust - Real neighborhoods, real problems, real solutions
- Stories stick - People remember stories 22x more than facts
- Proof prevents price shopping - They buy outcomes, not hours

The 3-Part Mini-Case Formula:

Part 1: Situation (The Setup)

Where + Who + Problem

- Name the neighborhood (not the specific address)
- Identify the decision maker type (HOA, PM, Municipality)
- State the specific challenge they faced

Part 2: Solution (Your Expertise)

What you recommended + Why it was different

- The specific change you made
- Why others missed it
- The expertise required

The 3-Part Mini-Case Formula:

Part 3: Result (The Proof)

Specific, measurable outcome

- Time saved (days, weeks)
- Money saved (percentage or dollar range)
- Trees preserved (number)
- Problems prevented.

The 3-Part Mini-Case Formula:

What Makes a Powerful Mini-Case:

- ■ Specific enough to be credible
- ■ Vague enough to protect client privacy
- ■ Includes numbers (time, money, or trees saved)
- ■ Shows expertise others might miss
- ■ Relates to common problems your market faces

Mini-Case Don'ts:

- ✗ Use client names without permission
- ✗ Use exact addresses
- ✗ Use identifiable photos
- ✗ Are unverifiable claims
- ✗ Competitor bashing

Your Mini-Case Builder Worksheet:

Recent Win to Document:

Situation: [Neighborhood] + [Decision Maker Type] faced [Specific Problem]

Solution: We [Specific Action] instead of [Common Approach]

Result: [Number] + [Outcome] + [Timeframe]

Example Fill-in:

Situation: Bayshore condos + HOA board faced 50% over budget on tree work

Solution: We prioritized by actual risk instead of treating all equally

Result: 35% under budget + addressed all high-risk trees + passed inspection

The 3-Part Mini-Case Formula:

The Psychology:

When you share specific mini-cases, property managers think:

- "They've solved this exact problem before"
- "They understand our constraints"
- "They measure results like we do"
- "They're worth calling"

PRO TIP!

Keep a notebook in your truck. After each win, jot down the numbers.

You'll forget the specifics in a week, but these mini-cases become your best sales tools.

Once you're sharing proof and real results, the next step is knowing what to measure. Not all metrics matter—here are the ones that actually move your business forward.

METRICS THAT MATTER

(Stop Counting Vanity Numbers)

The Uncomfortable Truth:

10,000 followers who don't need tree work = $0

50 local property managers who know your name = $500K pipeline

Most arborists track the wrong numbers because they feel good.

Let's track the numbers that buy new equipment.

Why These Metrics Matter:

LinkedIn rewards what it can measure.

But more importantly, YOU need to know if your 15 minutes daily are actually building **a business** or just building your **ego.**

The Only 5 Numbers That Count:

1 New LOCAL Connections (Weekly Target: 25)

Who: Property managers, HOA boards, municipal contacts IN YOUR SERVICE AREA

Why: Can't hire you if they don't know you

Track: Mark on calendar or simple spreadsheet

Red flag: Under 20/week means you're not consistent

The Only 5 Numbers That Count:

2 Total Saves on Your Posts (Weekly Target: 10+)

What: Total number of times your posts were bookmarked

Why: Proves your content is reference-worthy (people want to find it later)

Where to find: Click "View analytics" under each post → Look for "Saves"

What it means: Your content solves real problems worth keeping

Reality check: You can't see WHO saved, only how many

The Only 5 Numbers That Count:

3 Inbound Messages (Monthly Target: 4+)

What: "Hey, can you help with..." messages

Why: They're raising their hand without you asking

Quality check: Are they local? Are they decision makers?

Success indicator: Questions about your services, not "Great post!"

The Only 5 Numbers That Count:

4 Discovery Calls Booked (Monthly Target: 2+)

- **What**: Actual phone/video calls scheduled
- **Why**: Conversations convert to contracts
- **Source**: Track if they came from LinkedIn
- **Conversion goal**: 50% should move to proposal

The Only 5 Numbers That Count:

5 Proposals Accepted (The ONLY Metric That Pays Bills)

- **What**: Signed contracts from LinkedIn-sourced leads
- **Why**: Everything else is just activity
- **Track**: Revenue from LinkedIn vs. time invested
- **Goal**: One new recurring client per quarter

The Vanity Metrics to IGNORE:

✕ Total Follower Count

Unless they're local and qualified, *who cares*?

10K followers in India won't trim a single palm

✕ Viral Post Likes

That funny meme got 500 likes?
Cool.
How many were property managers?

Engagement from random people doesn't pay bills

The Vanity Metrics to IGNORE:

✕ Profile Views Without Context

100 views mean nothing if no one messages

WHO viewed matters more than how many

✕ Generic Comments

"Great post!"
"Thanks for sharing!"
= Zero value

Track meaningful conversations, not cheerleading

Now that you know which metrics matter, here's a simple weekly scorecard to help you track your consistency and momentum.

Your Weekly Scorecard:

Monday Morning (5 minutes): Check last week's numbers using this tracker

Introducing the **Connection-First Method**™ LinkedIn Weekly Success Tracker

This simple tracker helps you measure what truly matters — not just clicks or views, but meaningful actions that build visibility, credibility, and trust in your community. Use it every week to record your connections, saves, messages, and calls. At the end of 90 days, review your progress, highlight your wins, and identify the habits that are driving real growth.

Your Weekly Scorecard:

Monday Morning (5 minutes): Check last week's numbers using this tracker

Metric	Totals	Weekly Goal	Rationale
👫 Local Connections		25	Builds consistent local visibility (≈300 per 90 days)
■ Post Saves		10	Measures genuine engagement with your content (≈120 per 90 days)
💬 Inbound Messages		3	Indicates meaningful interest from prospects (≈12 per 90 days)
📞 Discovery Calls		1–2	Keeps a steady pipeline of qualified conversations (≈6 per 90 days)
🤝 Proposals Accepted		1	Reflects conversion success (≈3+ per 90 days)
■ Engagement Rate		2–5 %	Benchmark range for healthy post traction
🏆 Top-Performing Post		1	Identify which content drove the best reach or conversation
■ Notes / Reflections		–	Record what worked and what to adjust next week

Print This Page

Your Weekly Scorecard:

How to Use This Tracker:

- Print multiple copies or create a spreadsheet.
- Fill in every Monday morning
- Compare week-over-week progress
- Celebrate when you hit targets!

Your Weekly Scorecard:

What to Do With This Data:

- **If connections are low** → Not spending your 15 minutes daily
- **If saves are low** → Content isn't reference-worthy (add checklists, how-tos, cost breakdowns)
- **If messages are low** → Profile doesn't show expertise clearly
- **If calls are low** → Not making it easy to contact you
- **If proposals are low** → Wrong prospects or wrong conversation

Your Weekly Scorecard:

The 90-Day Reality Check:

- Month 1: Building foundation (connections + content)
- Month 2: Gaining traction (saves + messages)
- Month 3: Converting interest (calls + proposals)

How to find your saves:

1 Go to any of your posts and click 'View analytics' below it.

2 Look for the 'Saves' metric. On mobile, tap the post, then 'View analytics.'

3 Desktop users can also access analytics through their Activity dashboard.

A great post gets 5-10% of viewers saving it.

Your Weekly Scorecard:

The Math That Matters:

- 1,350 minutes invested (15 × 90 days)
- 300+ new local connections
- 36 valuable posts
- 8-12 discovery calls
- 2-4 new recurring clients
- ROI: One $30K annual contract = 1,333% return

Your Weekly Scorecard:

Simple Tracking Tool:

Create a note in your phone:

Week of [Date]:

Connections: ______________________________

Messages: ______________________________

Calls: ______________________________

Proposals: ______________________________

Won: ______________________________

The Bottom Line:
If it doesn't lead to local connections, conversations, or contracts, stop tracking it. Your time is worth more than vanity metrics.

With your actions and metrics in place, the next step is making sure your profile supports everything you're doing. Here's a clean, copy-ready profile tune-up you can plug in today.

PROFILE TUNE-UP (COPY-READY)

Banner (6-word promise)

- "Fewer Tree Headaches for HOAs"
- "Pass Tree Reviews Faster in [City]"
- "Safety-First Crews. Teach-to-Climb Path."

Headlines (choose one)

- Senior Arborist | Helping HOAs Cut Risk & Costs in [City]
- We Help Developers Pass Tree Reviews Faster ([County])
- Now Hiring Climbers | Train-to-Climb Path | [Your Company]

About (client focus — paste & tweak)

"We help HOA boards in **[City/County]** cut risk and unnecessary removals. Our reports are reviewer-friendly, our scopes prevent **lion's tailing**, and our goal is simple: safer trees, fewer surprises, lower lifetime costs."

About (recruiting focus — paste & tweak)

"We grow arborists. If you're new, you'll learn ground skills, knots, and safe rigging before you leave the ground.
If you're experienced, you'll run better gear, do cleaner work, and get home safe."

Featured

- Link your **HOA agenda checklist** (PDF or doc)
- Link your **Train-to-Climb 90-day path** (PDF)

With your profile tightened up, the next step is knowing what to say when people engage with your posts. These simple-scripts make those conversations easy.

TINY SCRIPT BANK

(What to Say AFTER They Engage)

Why This Matters:

The Connection-First Method™ says NO cold Direct Messages.

But what happens when someone likes your post, comments, or saves it? That's your green light to start a conversation.

These scripts help you turn their engagement into meaningful dialogue without being salesy.

The Golden Rule:

They engaged first = You've earned the right to reach out

Response Scripts (Copy, Paste, Personalize)

When they LIKE your post:

- "Thanks for the like on [topic]—want the 1-page printable version?"
- "Appreciate the thumbs up! If you're dealing with a similar [tree issue], I have a quick checklist that might help."
- "Hey [Name], saw you liked my post about [topic]. Happy to share the full guide if it's useful for your properties."

When they COMMENT on your post:

- "Great question—here's a quick visual. I can mark up your scenario if helpful."
- "[Name], that's exactly what most HOAs struggle with. Want me to send over the template we use?"
- "Thanks for adding that perspective! Based on your comment, this resource might save you time: [describe briefly]"

Why These Work

- You're responding to THEIR action (not cold pitching)
- You're offering additional value (not selling services)
- You're keeping it conversational (not corporate)
- You're making it easy to say yes or no
- Follow-Up Templates (For Engaged Connections)

Remember:

No engagement = No DM. Let your content do the work first.

TINY SCRIPT BANK
(WHAT TO SAY AFTER THEY ENGAGE)

Existing Client Exception

Important Note:
Current and past clients are the ONE exception to the "*No Engagement = No DM*" rule. You have an established business relationship, so reaching out is appropriate and expected.

Asking Existing Clients for Referrals:

> "Hi [Name], Really enjoyed working on [their property]. Quick question - do you know other property managers in [area] who might value our approach to tree care? Happy to be a resource for your network."

Why this exception makes sense:

- You've already provided value
- They know your work quality
- It's a natural business conversation
- They expect follow-up from vendors

PRO TIP!

Even better than a DM - create a post showcasing their project (with permission) and let them share it with their network naturally.

SAFETY, PERMISSIONS & ETHICS (TRUST CHECKLIST)

- ☑ Get photo permissions; blur identifying details; avoid addresses.
- ☑ Do not depict unsafe practices.
- ☑ Add an “education—not site-specific advice” footer on posts and PDFs.

First messages start the conversation. Follow-ups keep it moving. Use these templates to stay top-of-mind without feeling salesy.

FOLLOW-UP TEMPLATES

(For Engaged Connections)

After Estimate Follow-up

(They requested the estimate)**:**

> "Hi [Name], Thanks for letting me assess your trees yesterday. I've attached the proposal we discussed. Also sharing a post I wrote about [relevant topic from your discussion]. Let me know if you have questions!"

Re-engaging Dormant Connections

(Who previously engaged)**:**

> "Hey [Name], Noticed we haven't connected in a while. Just saw [something relevant to them - news, weather event, regulation change] and thought of your property. How are your trees handling [current season/situation]?"

As you start having real conversations on LinkedIn, it helps to understand the platform's language. These are the terms that will make everything feel more familiar.

LINKEDIN LINGO -
Speak the Language

1st/2nd/3rd Degree Connections

1st - People you're connected with

2nd - People connected to your connections (sweet spot!)

3rd - Everyone else

Engagement Rate

- (Likes + Comments + Shares + Saves) ÷ Impressions × 100
- Good rate: 2-5%

Saves (The Golden Metric)

- When someone bookmarks your post for later
- Indicates high-value, reference-worthy content

LINKEDIN LINGO -
Speak the Language

InMail

- Messages you can send to non-connections (Premium feature)
- You get 5/month with Premium

SSI (Social Selling Index)

- LinkedIn's 0-100 score of your effectiveness
- Check yours at: linkedin.com/sales/ssi

Impressions vs. Views

- Impressions: Times your post appeared on screen
- Views: People who actually looked at it

Creator Mode

- Special profile setting that emphasizes content
- Adds "Follow" button instead of "Connect"
- Not recommended for local relationship building

As you put everything into practice, a few common questions always come up. Here are the quick answers to keep you moving.

FREQUENTLY ASKED QUESTIONS

"What if I service multiple cities/counties?"

Create a primary location focus for your profile (your biggest market), but mention your full service area in your About section. In posts, rotate which areas you mention. Example: Monday's post mentions Tampa, Wednesday's mentions Clearwater, Friday's mentions St. Pete.

"How do I handle negative comments from competitors?"

Never engage in public arguments. If the comment is factually incorrect, provide correct information professionally. If it's just negative, either ignore it or respond with: "*Thanks for your perspective. Happy to discuss different approaches offline*." Then move on.

FREQUENTLY ASKED QUESTIONS

"Is LinkedIn Premium worth it?"

Not initially. Start with free LinkedIn for 90 days. Premium ($60/month) helps with: seeing who viewed your profile, 5 InMail credits per month, and better search filters. Consider it only after you're consistently posting and connecting. Most arborists succeed without ever upgrading.

"Can I use my company page instead of a personal profile?"

No. LinkedIn rewards personal profiles with **10x more reach** than company pages. Use your personal profile as the main strategy, then create a company page later as a credibility marker. People connect with people, not logos.

FREQUENTLY ASKED QUESTIONS

"What if property managers in my area aren't on LinkedIn?"

They're there, you just haven't found them yet. Try searching: facilities managers, community association managers (CAM), real estate professionals, and board members. Also, connect with vendors who serve them (landscapers, roofers, painters) as they share the same clients.

“What if I'm in a rural area with fewer properties?"

- Focus on a larger service radius
- Target neighboring towns
- Emphasize your willingness to travel

With the common questions out of the way, let's highlight a few LinkedIn features that can make your visibility and consistency even easier.

LINKEDIN FEATURES WORTH USING

Polls - Great for easy engagement

"What's your biggest tree concern this season?"	"How often does your HOA inspect trees?"	Run monthly, promotes discussion

LinkedIn Events - For local education

"Hurricane Prep for Property Managers"	"Free Tree Risk Assessment Workshop"	Virtual or in-person, builds authority

LINKEDIN FEATURES WORTH USING

Articles vs Posts

Posts: **200-1200 characters, daily visibility**	**Articles:** **Long-form content, less reach but more authority**	**Stick to posts until you have 1000+ connections**

LinkedIn Live - Advanced strategy

Requires application or approval	**Great for Q&A sessions**	**Partner with local organizations**

Even with the right tools and features, things won't always go smoothly. When that happens, here's how to recover quickly and get back on track.

WHEN THINGS GO WRONG (And How to Fix It) 🛠️

The Reality Check Nobody Talks About

You followed the guide. You fixed your profile. You sent connections. And... crickets.

Note: Some see results faster, especially in smaller markets or niche specialties

Before you blame the method (or yourself), know this: Every successful LinkedIn user hit these same walls. The difference? They knew how to troubleshoot.

If LinkedIn were easy, every arborist would have a full pipeline. Debugging is part of the process.

WHEN THINGS GO WRONG (And How to Fix It) 🛠

Problem #1: "I Sent 50 Connections, Only 2 Accepted"

Diagnosis Questions:

1. Does your profile clearly show what you do locally?
2. Are these people active on LinkedIn? (Check their last post date)

The Fix:

■ Immediate Actions:

Check your headline - Does it say WHO you help and WHERE?

Bad: "ISA Certified Arborist"

Good: "Helping Tampa HOAs Reduce Tree Risk & Costs"

WHEN THINGS GO WRONG (And How to Fix It) 🛠

Problem #1: "I Sent 50 Connections, Only 2 Accepted"

The Fix:

■ Immediate Actions:

Audit your targets - Are they:

- Active in the last 30 days?
- Are they really local to your service area?
- Connect with people within a 25 or 50-mile radius if possible.
- While that might be too far for you to go, those people probably have family or friends IN your service area that they can refer.

Yes, LinkedIn can be your referral partner too!

■ **Benchmark: 20-30% acceptance rate is normal.**

Below 10%? Your profile probably needs work.

WARNING: "Never send more than 20 connections per day. LinkedIn throttles aggressive connecting."

WHEN THINGS GO WRONG (And How to Fix It) 🛠

Problem #2: "No One Engages With My Posts"

The Brutal Truth:

Your posts are probably:

- Too technical for property managers
- Too generic (could be posted by any arborist anywhere)
- Missing a clear point or takeaway
- Posted at the wrong time

WHEN THINGS GO WRONG (And How to Fix It) 🛠

Problem #2: "No One Engages With My Posts"

The Fix:

The Engagement Audit: Look at your last five posts. Count how many include:

- A specific neighborhood or city name
- A number or statistic
- A clear lesson or takeaway
- A question at the end
- An image that isn't just a tree, a chainsaw, or someone cutting a tree (generic photo).

WHEN THINGS GO WRONG (And How to Fix It) 🛠️

Problem #2: "No One Engages With My Posts"

Score:

0-5:	6-15:	16-25:
You're posting for yourself, not your audience	You're on the right track, need refi nement	Your content should be converting

Example scoring:

- Post 1: Has city name + question = 2 points
- Post 2: Has statistic + lesson + good image = 3 points
- Post 3: Has nothing = 0 points
- Post 4: Has city name + lesson + question = 3 points
- Post 5: Has all five elements = 5 points
- Total: 13 points = "On the right track"

WHEN THINGS GO WRONG (And How to Fix It) 🛠

Problem #2: "No One Engages With My Posts"

The Fix - The Local Hook Formula:

[Local reference] + [Specific problem] + [Clear solution] + [Proof] + [Question]

Example:

"After last week's storms in Westshore, I noticed several HOAs dealing with damaged oaks from old pruning wounds.
Bayshore Gardens avoided this - we switched their specs from lion's tailing to proper reduction cuts 2 years ago.

Result:

15 mature oaks stood strong while neighbors lost trees. What pruning standards does your HOA follow?"

WHEN THINGS GO WRONG (And How to Fix It) 🛠

Problem #2: "No One Engages With My Posts"

Other Examples:

For Recruiting:

"Tampa tree companies are all fighting for the same 10 experienced climbers. We started our Train-to-Climb program last year instead.

Result:

5 ground guys became solid climbers who actually stick around. How are you building your next generation?"

For Municipal Work:

"Clearwater's new tree ordinance has property managers confused about the 60% canopy rule. We created a simple visual guide that helped three communities pass inspection on the first try. The reviewer even asked for copies. Need one for your next submittal?"

WHEN THINGS GO WRONG (And How to Fix It) 🛠

Problem #2: "No One Engages With My Posts"

The Fix:

■ Timing Test:

- Week 1: Post at 7 AM
- Week 2: Post at 12 PM
- Week 3: Post at 5 PM
- Track engagement and stick with what works

QUICK WIN:
"Add a local weather reference to any post for instant relevance. 'After last week's storms in [City]...'"

WHEN THINGS GO WRONG (And How to Fix It) 🛠

Problem #3: "I Got Leads But They Ghosted"

What Really Happened:

They were interested enough to reach out but not convinced enough to follow through. This is a trust issue, not a service issue.

The Fix:

■ The 48-Hour Rule:

When someone engages (not connects), you have 48 hours to build trust before they forget you exist.

- ➤ Hour 1-2: Thank them for engaging (comment or DM)
- ➤ Hour 24: Share something valuable (not a pitch)
- ➤ Hour 48: Ask a qualifying question

WHEN THINGS GO WRONG (And How to Fix It) 🛠

Problem #3: "I Got Leads But They Ghosted"

The Fix: The Value-First Response: When they say: "Tell me about your services"

❌ Don't say: "*We do pruning, removals, assessments...*"

✔ Do say: "*Happy to help! First, what tree challenges are you dealing with at [their property]? I have a quick checklist that might help identify priorities.*"

The Ghost Prevention Kit: If they aren't a connection and engaging with your post, connect with them on LinkedIn immediately! 👻

WHEN THINGS GO WRONG (And How to Fix It) 🛠

Problem #3: "I Got Leads But They Ghosted"

The Fix: If they are connected with you:

- Send a voice note (30 seconds max) - it's personal and memorable
- Share a relevant post you wrote and say, "*I thought this might help with what we discussed.*"
- Set a clear next step: "*I'll send over that HOA tree checklist Tuesday - sound good?*"

"The Follow-Up That Works"
"Hi [Name], I noticed a lot of HOAs in [Neighborhood] dealing with [specific issue] after our conversation. I put together a quick 1-pager on how others are handling it. Want me to send it over?"

WHEN THINGS GO WRONG (And How to Fix It) 🛠

Problem #4: "My Competitor Has 10x More Followers"

The Reality Check:

Follower count is vanity. Connection quality is sanity. If my target market were HOAs, I'd rather have 20 local property managers than 2,000 random arborists from around the world.

The Fix:

Stop Counting, Start Connecting:

Their 5,000 followers = 50 local decision makers	v.s.	Your 500 connections = 200 local decision makers

Who's winning?

WHEN THINGS GO WRONG (And How to Fix It) 🛠

Problem #4: "My Competitor Has 10x More Followers"

The Fix - The Competitor Advantage Play:

- Look at their popular posts
- See what questions people ask in comments
- Create better content answering those questions
- Connect with the people asking questions

WHEN THINGS GO WRONG (And How to Fix It) 🛠

Problem #4: "My Competitor Has 10x More Followers"

The Fix: The Authority Hack:

Instead of more followers, become the person people tag:

- Answer technical questions thoroughly
- Create reference content that people save
- Share municipal updates first
- Be helpful without pitching

"An arborist with 300 connections out-earned a competitor with 3,000 by focusing on local property managers exclusively. Quality over Quantity"

WHEN THINGS GO WRONG (And How to Fix It) 🛠

Problem #5: "I Don't Have Time for This"

The Time Audit Reality - Track your time for one day. You'll find:

- 2 hours on social media (yes, really - check your screen time)
- 30 minutes on random websites.
- 20 minutes explaining the same thing to customers.
- 15 minutes waiting for crews/clients.

That's 185 minutes you could redirect.

WHEN THINGS GO WRONG (And How to Fix It) 🛠

Problem #5: "I Don't Have Time for This"

The Fix - The Minimum Viable Routine:

- Monday (5 min): Send 5 connections while coffee brews
- Wednesday (5 min): Post one tip with a job photo
- Friday (5 min): Reply to any comments/messages

That's it. 15 minutes per day is all you need.

The Batch Method:

- Sunday: Write 3 posts for the week (30-45 min until you get faster). This is what I often do.
- Daily: Copy, paste, post (2 min)
- Friday: Send 20 connections for the week (10 min)

WHEN THINGS GO WRONG (And How to Fix It) 🛠️

Problem #5: "I Don't Have Time for This"

The Fix - The Integration Method:

- Waiting for the client? Send connection requests.
- Crew loading up? Write down or dictate a quick post idea. Not comfortable with voice notes? A personalized video using Loom or even a well-crafted text message works too.
- Lunch break? Reply to comments
- Evening TV? Schedule tomorrow's post

REALITY CHECK:

You spent 2 hours on that estimate that didn't close. LinkedIn takes 15 minutes and can build your pipeline.

WHEN THINGS GO WRONG (And How to Fix It) 🛠️

Problem #6: "LinkedIn Limited My Connections"

The Warning Signs:

- ⚠️ "You've reached the weekly invitation limit" message
- ⚠️ Can't send new connection requests
- ⚠️ "Pending" invitations stuck for weeks
- ⚠️ Account feels restricted

Why This Happens: LinkedIn has weekly limits to prevent spam

- ➤ Free accounts: ~100-200 connections/week
- ➤ Too many pending invitations (300+)
- ➤ Too many "I don't know this person" responses

WHEN THINGS GO WRONG (And How to Fix It) 🛠

Problem #6: "LinkedIn Limited My Connections"

The Fix - Immediate Recovery:

- Stop ALL connection activity for 7 days
- Withdraw old pending invitations (My Network → Manage → Sent → Withdraw)
- Focus on engaging with existing connections.
- Let people connect with YOU instead

WHEN THINGS GO WRONG (And How to Fix It) 🛠

Problem #6: "LinkedIn Limited My Connections"

The Fix - Prevention Going Forward:

- Never exceed 20 connections/day
- Spread requests throughout the week
- Only connect with 2nd-degree connections (higher acceptance rate)
- Withdraw pending invitations older than 2 weeks
- Track your weekly totals

PRO TIP:
If limited, use the time to create killer content that makes people want to connect with you.

WHEN THINGS GO WRONG (And How to Fix It) 🛠

Problem #7: "My Posts Get Zero Reach Now"

You Might Be Shadowbanned If:

- Engagement dropped 90% overnight
- Only 10-20 views per post (all connections)
- Hashtags don't increase reach
- New connections don't see your content

Common Shadowban Triggers:

- Using banned hashtags (#followforfollow, #l4l)
- Posting the same content repeatedly
- Too many external links
- Automation tool detection
- Reported for spam by multiple users

Problem #7: "My Posts Get Zero Reach Now"

The Recovery Plan

Week 1 - Full Stop:

- Stop posting for 48-72 hours
- Remove ALL automation tools
- Delete any duplicate posts
- Review LinkedIn's user agreement

Week 2 - Slow Restart:

- Post 100% original content only
- No external links for 2 weeks
- Focus on thoughtful comments (10+ daily)
- Use NO hashtags initially
- One post every other day, maximum

Week 3+ - Gradual Return:

- Slowly add 1-2 relevant hashtags
- Continue high-value engagement
- Monitor view counts carefully
- Build back to normal posting schedule

Timeline:
Most shadowbans lift within 14-30 days if you follow the recovery plan.

WHEN THINGS GO WRONG (And How to Fix It) 🛠

Problem #8: "My Connection Requests Get Rejected"

High Rejection Indicators:

- Multiple "I don't know this person" responses
- Acceptance rate below 20%
- LinkedIn warning messages

The Fix - Improve Your Targeting

Only connect with people who:

- Are in your geographic area
- Share mutual connections
- Are active (posted in the last 30 days)
- Match your ideal client profile

WHEN THINGS GO WRONG (And How to Fix It) 🛠

The Fix - Make Yourself More Connectable

Ensure your profile clearly shows:

- Your location
- What you do
- Who you help
- Professional photo
- Post valuable content before connecting
- Engage with their content first

REMEMBER:
Quality > Quantity.
50 perfect connections beat 500 random ones.

WHEN THINGS GO WRONG (And How to Fix It) 🛠

The Debug Checklist

When something isn't working, run through this:

Profile:

- Your headline mentions your location and your value statement.
- Photo shows you as a professional (Showing off your cool new gear instead of being client-focused).
- The About section speaks to clients, not peers.
- Featured section has valuable content.

Connections:

- Target 2nd degree connections
- Just click connect - no notes
- Connect with active users
- Track acceptance rates

The Debug Checklist

When something isn't working, run through this:

Content:

- Your posts include local references
- You share knowledge, not just services (educate people).
- Ask questions to encourage engagement
- Respond to every comment

Follow-Through:

- Respond within 24 hours
- Provide value before pitching
- Set clear next steps
- Follow up consistently

WHEN THINGS GO WRONG (And How to Fix It) 🛠

Remember: LinkedIn is a Long Game

- Week 1-2: Feels like shouting into the void
- Week 3-4: A few people notice
- Week 5-8: Engagement picks up
- Week 9-12: Inbound leads start flowing
- Month 4+: You can't imagine not using it

The winners aren't the ones who never struggled. They're the ones who troubleshoot and keep going.

WHEN THINGS GO WRONG (And How to Fix It) 🛠️

What Not To Do:

- Don't buy connections
- Don't use automation tools
- Don't copy/paste the same message 50 times
- Don't connect and immediately pitch

Quick Fix Cheat Sheet

- *Low Acceptance Rate?* → Fix your headline
- *No Engagement?* → Add local references
- *Leads Ghosting?* → Connect immediately when they engage
- *Feeling Behind?* → Focus on quality connections, not follower count
- *No Time?* → 15 minutes daily minimum

Every 'overnight success' on LinkedIn spent their first month talking to themselves. Push through!

The Connection-First Method™ takes skill and consistency. AI helps you do both with less effort. Here are the tools that actually make a difference.

WORKING SMARTER WITH AI TOOLS

Your Secret Weapon for Content Creation

The Reality:

You're an arborist, not a copywriter. You know trees, not hashtags. That's where AI can become your content partner, if you use it correctly.

AI doesn't replace your expertise - it amplifies your voice

What AI Can Do For You

Turn Your Expertise into Posts

You say: "Oak wilt is bad in spring."

AI helps: *"Property managers: That oak looking thin? Spring oak wilt can spread to every oak on your property within weeks. Here's what to check today..."*

WORKING SMARTER WITH AI TOOLS

What AI Can Do For You

Fix Your "Arborist Brain" Writing

You write: "Conduct crown reduction per ANSI A300 standards."

AI translates: "We'll carefully reduce your tree's height while keeping it healthy and beautiful"

Generate Ideas When You're Stuck

Input: Your recent jobs, local weather, and season

Output: 10 post ideas specific to your market

Create Templates from Your Best Content

Feed it your highest-performing post

Get 5 variations you can use next month

WORKING SMARTER WITH AI TOOLS

My Recommended AI Setup

There are many AI tools available. After testing every AI tool out there, I use Stanley (https://stanley.stan.store) for all my LinkedIn content. It's built specifically for LinkedIn creators and understands the platform better than generic AI tools. While I use and recommend Stanley, any AI tool that helps you create better content faster is worth exploring.

Stanley helps me:

- Write posts that sound like ME, not a robot
- Generate ideas based on what's working NOW
- Keep my content consistent when I'm busy
- Track what resonates with my audience

If you want to support this guide and try Stanley out, just click the link.

WORKING SMARTER WITH AI TOOLS

The Right Way to Use AI

"AI is your co-pilot, not your pilot"

Do:	Don't
Start with YOUR stories and expertise	Copy-paste without reading
Use AI to polish and structure	Let AI make up tree facts
Always fact-check technical details	Use generic business advice
Add your local specifics	Forget your unique voice
Review and personalize before posting	Post AI images of trees **(everyone can tell)**

WORKING SMARTER WITH AI TOOLS

The Right Way to Use AI

Quick AI Workflow

1. Brain dump your idea (messy is fine)
   ↓
2. Feed to AI with context: "I'm an arborist in [City] writing for property managers"
   ↓
3. Refine the output with your specifics
   ↓
4. Add local details, regulations, and examples
   ↓
5. Review for accuracy and voice
   ↓
6. Post with confidence

WORKING SMARTER WITH AI TOOLS

The Right Way to Use AI

Example Prompts That Work

COPY THESE EXACT PROMPTS

For Property Managers:
"Write a LinkedIn post about [TOPIC] for property managers in [CITY]. Include a specific local example and end with a question. Keep it under 200 words."

For Recruiting:
"Create a job posting that would appeal to young arborists. Emphasize safety, training, and career growth. Make it sound exciting, not corporate."

For Authority Building:
"Turn this technical tree care concept into a LinkedIn post that municipal arborists would save: [INSERT CONCEPT]. Include why it matters for urban forestry."

WORKING SMARTER WITH AI TOOLS

The Right Way to Use AI

Beyond Writing: Other AI Uses

- Image descriptions: For accessibility
- Comment responses: When you need ideas for thoughtful replies
- Video scripts: For LinkedIn video content
- Email templates: For following up with connections
- Content calendars: Planning your monthly themes

5 hours of content creation becomes 30 minutes with the right AI workflow

The Integration Rule

AI + Your Expertise + Local Knowledge = Content That Converts

Remember: Property managers can smell generic content from a mile away. AI helps you package your expertise, not replace it.

THE COMPETITION

The Connection-First Method™ works best when you know exactly how you're different from everyone else.

Let's talk about the competition and why most arborists never stand out.

The Competition Reality Check

First, Let's Address the Elephant in the Room

You're scrolling LinkedIn and see it: Your biggest competitor just posted about the same municipal tree ordinance you were planning to write about. Your stomach drops. Mine did when I saw it.

THE COMPETITION

Here's what's actually happening:

- They have 47 likes (45 are other arborists)
- 2 property managers saw it
- 0 property managers saved it
- **You're worrying about nothing**

The Scarcity Mindset (What Keeps You Broke)

- "If they get that HOA, I won't."
- "I can't share my knowledge - it's my edge." ***Spoiler:*** *We all know the same techniques. The difference is execution.*
- "They have more followers, so why bother?"
- "I should copy what's working for them."

THE COMPETITION

When You See Competitors on LinkedIn

Reality Check: In a 10-mile radius of your business, there are probably:

- 50+ HOA communities
- 1,000+ commercial properties
- 50,000+ homes with trees
- 5-10 tree services actually using LinkedIn

You're fighting over 0.01% of the market while ignoring 99.99%.

THE COMPETITION

When You See Competitors on LinkedIn

The Abundance Approach (What Makes You Money)

✓ **Study Their Engagement Patterns**

- Which posts get property managers commenting?
- What questions appear repeatedly?
- What topics get saved vs. just liked?

✓ **Mine Their Connections**

- View their connections list
- Filter for property managers, HOA board members
- Connect with mutual interests: "Saw we both know [Name]..."

THE COMPETITION

When You See Competitors on LinkedIn

✓ **Differentiate Your Expertise**

If They Post About...	You Post About...
Big removals	Preservation techniques
Equipment porn	Safety procedures
Storm damage	Storm prevention
Price/discounts	Value/ROI
Generic tree facts	Local ordinance specifics

THE COMPETITION

When You See Competitors on LinkedIn

The Competitor Content Audit

Your Weekly 10-Minute Competitor's Audit:

1. Check 3 local competitors' profiles
2. Note their last 3 posts
3. Identify gaps they're not covering
4. Plan content that fills those gaps

THE COMPETITION

When You See Competitors on LinkedIn

What to Look For:

- Topics they avoid (usually compliance/ technical stuff)
- Questions in their comments, they don't answer well
- Connection requests they're ignoring
- Their posting frequency (opportunity to be more consistent)

Strategic Engagement **(The High Road Wins)**

THE COMPETITION

When You See Competitors on LinkedIn

When Competitors Post Good Content:

✓ Like it (shows confidence)

✓ Add valuable perspective: "Great point about oak wilt. In [County], we also need to consider..."

✓ Tag relevant people: "John Smith thought you'd fi nd this useful for Oakwood HOA"

Why This Works:

- Property managers see you as collaborative, not cutthroat
- You appear confident and established
- Their audience becomes aware of you
- LinkedIn algorithm shows you to their network

The "Competitor Collaboration" Play

THE COMPETITION

When You See Competitors on LinkedIn

Sometimes the best move is unexpected:

"Hey [Competitor], noticed we're both educating property managers about the new ordinance. Want to co-host a LinkedIn Live to answer questions? We could cover different angles - you handle residential, I'll handle commercial."

What happens:

- Both audiences grow
- Property managers see professionalism
- You establish market leadership together
- Referrals start flowing both ways

It will happen. Here's your response:

1. Take it as a compliment - You're worth copying
2. Stay ahead - They can copy yesterday's content, not tomorrow's insights
3. *Go deeper* - If they copy your "5 pruning tips," you post "The science behind why these 5 tips work."
4. Build relationships - They can't copy your connections

THE COMPETITION

When They Copy You

The Ultimate Competition Strategy

Stop competing on their terms. Create your own game:

- They fight for the lowest price? You educate on value.
- They post removal photos? You post preservation wins.
- They chase everyone? You attract the right ones.
- They copy content? You build relationships.

Remember: The arborist who educates the market grows the market. Rising tides lift all boats - except the ones with holes (bad operators naturally fail).

THE COMPETITION

Your "Competitor-Proof" Content Strategy

- *Monday:* Educational content they won't copy (too much work)
- *Wednesday:* Local relationship building (can't be copied)
- *Friday:* Behind-the-scenes culture (uniquely yours)

The 90-Day Result:

While they're watching and copying you, you've built relationships with 100 property managers who see you as the educator and them as the imitator.

Standing out is important, but staying visible all year is what keeps you top-of-mind. This seasonal calendar makes your posting easy and predictable.

SEASONAL CONTENT CALENDAR

Year-Round Content That Works

Spring (March-May)

- Storm prep inspections
- Post-winter damage assessments
- Budget planning for summer work
- Growth regulation timing

Summer (June-August)

- Heat stress identification
- Watering myths vs. reality
- Lightning protection systems
- Why summer pruning can be wrong

SEASONAL CONTENT CALENDAR

Year-Round Content That Works

Fall (September-November)

- Hurricane prep (Gulf states)
- Pre-winter inspections
- Budget season for HOAs
- Planting season education

Winter (December-February)

- Dormant season pruning benefits
- Annual contract planning
- Ice damage prevention
- Certification/training season

Year-Round Winners:

- Local ordinance updates
- "What this week's weather means for your trees"
- Board meeting prep tips
- Safety moments from job sites

Now that you have the full system, here are the tools and resources that will help you keep going.

RESOURCES & LINKS

LinkedIn Help & Best Practices

- LinkedIn User Agreement: linkedin.com/legal/user-agreement
- LinkedIn Help Center: linkedin.com/help
- Professional Community Policies: https://www.linkedin.com/help/linkedin/answer/a403269/
- LinkedIn Learning (free courses): linkedin.com/learning

Tools Mentioned in This Guide

- Canva (free banner/graphic creation): canva.com
- Stanley (AI LinkedIn assistant): getstanley.com
- Loom (video messages): loom.com
- LinkedIn Premium (optional upgrade): linkedin.com/premium

RESOURCES & LINKS

Connect With Me

LinkedIn: linkedin.com/in/johnwsnow

Company: Tree CheckUp, LLC

Email: JohnSnow@Treecheckup.com

Website: www.ArboristMentoring.com

Join the Smart Tree Care Community

Connect with other growth-minded arborists implementing these strategies. Share wins, ask questions, and learn from each other's experiences. Join here:

https://treecheckuppro.treecheckup.com/checkout/pro-arborists

RESOURCES & LINKS

Stay Updated

This guide is updated annually as LinkedIn evolves.

To get notified of updates:

- Follow me on LinkedIn
- Comment "UPDATE ME" on any of my posts
- Save this guide and check back

Need Help?

- For LinkedIn technical issues: linkedin.com/help
- For strategy questions: DM me on LinkedIn

WANT TO GO DEEPER?

Option 1: 1:1 Coaching

Your 30-Minute Strategy Session - $90

Stuck on something specific? Let's talk through your exact situation:

- Review your profile together
- Identify your best opportunities
- Create your custom 30-day plan
- Get answers to your specific questions

Book Your Strategy Call

https://treecheckup.com/linkedin-consultation

WANT TO GO DEEPER?

Option 2: The 4-Week LinkedIn Challenge - $297

Join other arborists implementing this system together:

- Weekly 60-minute group Zoom calls
- Assignments with accountability
- Direct access to me for questions
- Small group (max 20 arborists)
- Next cohort starts: January 3rd, 2026

Reader Special: Save $50 with code GUIDE50

Sign up to get notified about the next LinkedIn Challenge Cohort

https://stan.store/JohnSnow/p/linkedin-challenge-cohort

TAKE THE QUIZ

You've just worked through the entire Connection-First Method™, from fixing your profile to building real visibility, trust, and conversations on LinkedIn. The quiz isn't here to trick you or test your memory. It's here to help you sanity-check your understanding of the core ideas that actually drive results. If you can answer these confidently, you're not just reading, you're ready to implement.

This quiz focuses on the habits, principles, and decisions that separate arborists who get results from those who stay invisible.

Take the Quiz

https://arboristmentoring.com/linkedin-quiz

TREE CHECK UP

YOUR SUCCESS IS MY SUCCESS

Made It This Far? You're Already Ahead.

Most arborists bookmark guides like this and never open them again. You're different. You're here at the end, which means you're serious about growing your business, career, or both.

You don't need to implement everything at once. Start with one track, one posting rhythm, and one weekly win. The arborists who win on LinkedIn aren't the loudest—they're the most consistent.

Here's what I know:

In 90 days, you'll either be glad you started today, or you'll wish you had.

Share Your Wins

Nothing makes me happier than hearing success stories from arborists who took action. When you land that first LinkedIn client, get that perfect hire, or become the go-to expert in your market, I want to know about it.

DM me your success story on LinkedIn:

- How many connections did you start with
- Which track did you choose
- Your first win (no matter how small)
- What surprised you most

With permission, I feature success stories to inspire other arborists who are in the same place as you, ready to start, but needing that extra push.

One Last Thing...

Remember that example from the introduction? The arborist who posted tree content that only other arborists would see?

I see this every day. Great arborists with valuable knowledge, wasting it on the wrong audience. They're talking to each other instead of talking to clients.

Today, property managers in three states know my name. Municipalities call me for ordinance reviews. My network sends me more work than I can handle.

The difference? I learned to use LinkedIn **strategically** instead of socially. To connect with clients instead of competitors. To share the value that clients actually save.

Your turn starts now.
See you on LinkedIn,

John W. Snow

P.S. Still skeptical? Give it 30 honest days. Follow the system. Track your results. If it doesn't work, DM me and tell me why. I'll personally help you troubleshoot. That's how confident I am in this method.

TREE CHECK UP

ISBN: 979-8-218-92198-9

This book is provided for informational and educational purposes only. The strategies, examples, and opinions expressed are based on the author's experience and are not a guarantee of specific business, career, or financial outcomes. Results will vary depending on market conditions, consistency, implementation, and individual circumstances.

The Connection-First Method™ is a trademark of Tree CheckUp, LLC and ArboristMentoring.com. All other trademarks referenced belong to their respective owners.

LinkedIn is a registered trademark of LinkedIn Corporation. This book is not affiliated with, endorsed by, or sponsored by LinkedIn.

The author and publisher disclaim any liability for actions taken or not taken based on the contents of this book. Readers are responsible for ensuring compliance with all applicable laws, regulations, professional standards, and platform policies.

www.ingramcontent.com/pod-product-compliance
Lightning Source LLC
LaVergne TN
LVHW060035210926
848481LV00013B/3872